INTRODUCTION

Penicillin is a group of antibiotics (medicinal drugs used to treat bacterial infections) that work in a particular way to damage bacteria on your frame. Types of penicillin and capsules carefully related to them are known as "penicillins." They're a subclass of beta-lactam antibiotics. Penicillin antibiotics are available in oral form (tablets or liquid you swallow) and IV form (liquid that an issuer injects at once into your vein).

WHAT ARE THE SORTS OF PENICILLIN?

Two of the main differences among the varieties of penicillin are the way they're made and what form of micro organism they're powerful in opposition to. Types of penicillin encompass:

- Natural penicillins. Natural penicillins are determined within the world around us (they aren't made by way of humans). Scientists isolate (purify) just the penicillin from wherein it's located to make it into medicines.

- Semi-synthetic penicillins. Scientists alter penicillin's herbal form to make extra effective antibiotics. These are called semi-artificial penicillins. They encompass penicillinase-resistant penicillins, aminopenicillins and prolonged-spectrum penicillins.

- Combination medicinal drugs. Penicillins are frequently combined with different medications into one drug to help them paintings higher.

SEMI-SYNTHETIC PENICILLINS

Semi-artificial penicillins consist of:

•   Penicillinase-resistant penicillins. Nafcillin, oxacillin and dicloxacillin are penicillinase-resistant penicillins. They are available each IV and pill shape and are frequently used to treat staph infections. Methicillin, that is not often used anymore, is also a penicillinase-resistant penicillin.

•   Aminopenicillins. Amoxicillin and ampicillin are aminopenicillins. Amoxicillin is one of the most commonly used penicillins. Providers use it to deal with ear infections, UTIs, pneumonia and different not unusual infections. It is available in a pill or liquid you can swallow. Ampicillin comes in tablet or IV shape.

- Extended-spectrum penicillins. Piperacillin is extended-spectrum penicillin. Providers use it for hard-to-deal with infections, like Pseudomonas aeruginosa. Other extended-spectrum penicillins, like carbenicillin and ticarcillin, are discontinued within the U.S.

HOW DOES PENICILLIN PAINTINGS?

Penicillin works with the aid of attaching to the wall of micro organism cells. It damages the cell wall and ultimately destroys the bacteria. Over time, micro organism has evolved resistance to herbal penicillins. This method that certain styles of bacteria can save you antibiotics from detrimental their cells. This has befallen commonly considering that penicillin's first use. Each time, scientists evolved new antibiotics that worked inside the same methods as herbal penicillin — via attaching to the cellular wall

— however had strategies for warding off antibiotic resistance. Some bacteria, like MRSA (methicillin-resistant Staphylococcus aureus), withstand all types of penicillins. This can make them very tough to treat. Scientists have developed antibiotics that paintings in extraordinary ways to try and treat those infections.

OTHER MEDICAL PROBLEMS

The presence of different scientific troubles might also affect the use of medicines on this elegance. Make sure you tell your doctor when you have every other clinical problem, specially:

- Allergy, standard (inclusive of allergies, eczema, hay fever, and hives), records of—Patients with records of preferred allergies may be much more likely to have an extreme response to penicillins

- Bleeding problems, history of—Patients with records of bleeding issues may be more likely to have bleeding whilst receiving carbenicillin, piperacillin, or ticarcillin

- Congestive coronary heart failure (CHF) or

- High blood strain—Large doses of carbenicillin or ticarcillin may make those situations worse, because those drugs incorporate a large amount of salt

- Cystic fibrosis—Patients with cystic fibrosis can also have an extended risk of fever and skin rash whilst receiving piperacillin

- Kidney ailment—Patients with kidney disorder might also

have an accelerated danger of facet outcomes

- Mononucleosis ("mono")—Patients with mononucleosis may additionally have an extended chance of pores and skin rash whilst receiving ampicillin, bacampicillin, or pivampicillin

- Phenylketonuria—some strength of the amoxicillin chewable tablets comprises aspartame, which is modified by the body to phenylalanine, a substance that is dangerous to sufferers with phenylketonuria.

- Stomach or intestinal disease, history of (mainly colitis, consisting of colitis due to antibiotics)—Patients with a history of belly or intestinal ailment can be more likely to broaden colitis whilst taking penicillins

PROPER USE

Penicillins (except bacampicillin capsules, amoxicillin, penicillin V, pivampicillin, and pivmecillinam) are pleasant focused on a complete glass (8 oz.) of water on an empty stomach (both 1 hour before and 2 hours after food) except in any other case directed via your medical doctor. For sufferers taking amoxicillin, penicillin V, pivampicillin, and pivmecillinam:

- Amoxicillin, penicillin V, pivampicillin, and pivmecillinam can be taken on a complete or empty belly.

- The liquid form of amoxicillin will also be taken through itself or mixed with formulation, milk, fruit juice, water, ginger ale, or different cold drinks. If mixed with different drinks, take immediately after mixing. Be sure to drink all of the liquid to get the full dose of medication. For sufferers taking bacampicillin:

- The liquid form of this medication is pleasant all for a complete glass (eight oz) of water on an empty belly (either 1 hour earlier than or 2 hours after meals) except in any other case directed by way of your medical doctor.

- The pill shape of this medication may be taken on a complete or empty belly. For sufferers taking penicillin G by mouth:

- Do not drink acidic fruit juices (as an example, orange or grapefruit juice) or different acidic liquids within 1 hour of taking penicillin G since this can keep the medication from working well. For sufferers taking the oral liquid form of penicillins:

- This medication is to be taken through mouth even though it comes in a dropper bottle. If this remedy does now not come in a dropper bottle, use an in particular marked measuring spoon or different tool to degree each dose appropriately. The average household teaspoon may not preserve the right quantity of liquid.

- Do now not use after the expiration date on the label. The medicine may not work nicely after that date. If you've got any questions on this, check with your pharmacist. For sufferers taking the chewable pill form of amoxicillin:

- Tablets should be chewed or crushed earlier than they are

swallowed. To help clear up your contamination absolutely, keep taking this medicinal drug for the full time of treatment, even if you begin to experience higher after some days. If you've got a"strep" infection, you have to preserve taking this remedy for as a minimum 10 days. This is in particular vital in"strep" infections. Serious coronary heart troubles ought to expand later if your contamination isn't cleared up absolutely. Also, if you prevent taking this remedy too quickly, your signs may return. This medicinal drug works great when there may be a regular quantity in the blood or urine. To help keep the amount consistent, do no longer leave out any doses. Also, it is best to take the doses at calmly spaced instances, day and night time. For instance, if you are to take four doses an afternoon, the doses should be spaced about 6 hours apart. If this interferes with your sleep or other daily activities, or in case you need assist in planning the satisfactory times to take your medication, take a look at together with your health care expert. Make sure your health care professional knows in case you are on a low-sodium (low-salt) food regimen. Some of these medicines incorporate enough sodium to cause problems in some humans.

PRECAUTIONS

If your symptoms do now not enhance within a few days, or in the event that they turn out to be worse, test together with your physician. Penicillins can also purpose diarrhea in some sufferers.

- Check with your doctor if excessive diarrhea occurs. Severe diarrhea may be a sign of a serious side impact. Do now not take any diarrhea medication without first checking together with your doctor. Diarrhea medicines may additionally make your diarrhea worse or make it final longer.

- For slight diarrhea, diarrhea remedy containing kaolin or attapulgite (e.G., Kaopectate pills, Diasorb) can be taken. However, different kinds of diarrhea remedy should no longer be taken. They might also make your diarrhea worse or make it ultimate longer.

- If you have got any questions on this or if moderate diarrhea keeps or gets worse, check together with your fitness care professional. Oral contraceptives (delivery manage capsules) containing estrogen may not paintings properly if you take them whilst you're taking ampicillin, amoxicillin, or penicillin V. Unplanned pregnancies may additionally arise. You ought to use an exceptional or additional manner of beginning manage whilst you're taking any of those penicillins. If you have got any questions on this, take a look at together with your health care professional. For diabetic patients:

- Penicillins may additionally purpose fake check effects with some urine sugar checks. Check with your doctor before changing your weight-reduction plan or the dosage of your diabetes medicine.

Before you have got any medical tests, inform the physician in rate which you are taking this medicinal drug. The outcomes of some assessments may be stricken by this medicinal drug.

HOW HAVE TO I USE THIS MEDICINAL DRUG?

Take this drug via mouth. Take it as directed at the prescription label on the equal time every day. You can take it without or with meals. If it upsets your belly, take it with food. Take this entire drug until your health care issuer tells you to stop it early. Keep taking it even if you assume you're higher. Talk to your health care issuer approximately the use of this drug in children. While it may be prescribed for youngsters as younger as 12 for decided on conditions, precautions do apply.

Overdosage: If you watched you've got taken an excessive amount of of this medicinal drug touch poison control middle or emergency room right away.
NOTE: This medication is best for you. Do no longer percentage this medicinal drug with others.

WHAT SHOULD I WATCH FOR AT THE SAME TIME AS USING THIS MEDICINAL DRUG?

Tell your medical doctor or health care company if your signs and symptoms do now not improve. This medication may additionally reason serious skin reactions. They can appear weeks to months after starting the drugs. Contact your fitness care provider right away in case you are aware fevers or flu-like signs and symptoms with a rash. The rash can be red or purple and then change into blisters or peeling of the pores and skin. Or, you would possibly note a red rash with swelling of the face, lips or lymph nodes on your neck or below your palms. Do not treat diarrhea with over the counter products. Contact your physician when you have diarrhea that lasts more than 2 days or if it's miles severe and watery. If you have got diabetes, you could get a fake-superb result for sugar in your urine. Check along with your health practitioner or fitness care issuer. Birth manage pills might not paintings properly whilst you're taking this medicinal drug. Talk on your medical doctor about the use of a further technique of delivery control.

WHAT SIDE RESULTS MIGHT ALSO I NOTE FROM RECEIVING THIS MEDICINE?

Side results that you have to document on your physician or health care professional as soon as feasible:

- Allergies like skin rash or hives, swelling of the face, lips, or tongue

- Respiratory problems

- Fever

- New signs and symptoms of contamination

- Redness, blistering, peeling or loosening of the pores and skin, which include in the mouth

- Unusually susceptible or worn-out Side consequences that commonly do not require scientific attention (report in your doctor or fitness care expert in the event that they preserve or are bothersome):

- Diarrhea

- Headache

- Nausea, vomiting

- Sore mouth or tongue

- Belly disenchanted

This listing may not describe all possible aspect consequences. Call your health practitioner for scientific advice approximately side consequences. You can also report side outcomes to FDA at 1-800-FDA-1088.

WHAT ARE THE DIFFERENCES AMONG PENICILLINS?

The herbal penicillins (penicillin G and penicillin V) are simplest active towards gram-positive micro organism (see under for an explanation). Penicillin V is greater acid-resistant than penicillin G, this means that it can be taken orally. Modern semi-artificial penicillins include ampicillin, carbenicillin (discontinued), and oxacillin. These may be taken orally, have a few degree of resistance to beta lactamase, and are powerful towards some gram-terrible micro organism. Most micro organism can be classified as gram-fine or gram-negative based totally on variations of their cell wall structure, which may be prominent under a microscope the usage of a form of dye. One of the most vital variations between those forms of micro organism is that gram-high quality micro organism are extra liable to antibiotics at the same time as gram-negative micro organism are more proof against antibiotics. Antipseudomonal penicillins, which include piperacillin and ticarcillin (discontinued), are penicillins which have extra hobby towards a few hard-to-kill kinds of gram-terrible micro organism such as Pseudomonas, Enterococcus and Klebsiella. They are useful for urinary tract infections because of inclined bacteria due to the fact they pay attention in the urine. Some penicillins are combined with a beta-lactamase inhibitor. A beta-lactamase inhibitor blocks the activity of beta-lactamase enzymes however tends to have little antibiotic interest on its personal. Some penicillins (along with oxacillin, dicloxacillin, and nafcillin) are obviously proof against sure beta-lactamases and are known as penicillinase-resistant penicillins. Others, inclusive of amoxicillin, ampicillin, and piperacillin will have their interest extended by combining them with a beta-lactamase inhibitor. Clavulanate, sulbactam, and tazobactam are all beta-lactamase inhibitors.

WHAT SPECIAL PRECAUTIONS HAVE TO I OBSERVE?

Before taking penicillin V potassium,

- inform your health practitioner and pharmacist in case you are allergic to penicillin V potassium, different penicillin antibiotics, cephalosporin antibiotics along with cefaclor, cefadroxil, cefazolin (Ancef, Kefzol), cefepime (Maxipime), cefixime (Suprax), cefotaxime (Claforan), cefotetan, cefoxitin (Mefoxin), cefpodoxime, cefprozil, ceftaroline (Teflaro), ceftazidime (Fortaz, Tazicef, in Avycaz), ceftibuten, ceftriaxone, cefuroxime (Ceftin, Zinacef), and cephalexin (Keflex); any other medicines, or any of the components in penicillin V potassium capsules or oral solution.

• tell your doctor and pharmacist what prescription and nonprescription medicinal drugs, vitamins, dietary dietary supplements, and herbal products you are taking or plan to take. Your doctor might also need to change the doses of your medications or screen you carefully for side consequences.

- tell your health practitioner in case you currently have nausea or vomiting. Also, tell your medical doctor when you have or have ever had allergies, allergies, hay fever, or kidney disease.

- tell your health practitioner in case you are pregnant, plan to come to be pregnant, or are breastfeeding. If you end up pregnant even as taking penicillin V potassium, name your physician.

HOW OUGHT TO THIS MEDICATION BE USED?

Penicillin G injection comes as a powder to be mixed with water and as a premixed product. Penicillin G injection is normally injected right into a muscle or vein but can also be given immediately into the lining of the chest hollow space, into the fluid surrounding the spinal wire, or into a joint or different regions. The quantity of doses you receive each day and the overall period of your treatment rely on your preferred fitness, the form of contamination which you have, and how well you reply to the medicine.

You might also acquire penicillin G injection in a clinic or you may administer the drugs at domestic. If you may be receiving penicillin G injection at home, your healthcare issuer will show you the way to use the drugs. Be positive which you recognize those instructions, and ask your healthcare issuer when you have any questions. You must start to experience better throughout the primary few days of remedy with penicillin G injection. If your signs and symptoms do no longer enhance or get worse, call your medical doctor.

Use penicillin G injection for as long as your health practitioner tells you that you have to, even if you feel better. If you prevent the usage of penicillin G injection too soon or skip doses, your contamination won't be absolutely handled and the micro organism might also turn out to be resistant to antibiotics. If you're the usage of penicillin G injection to treat certain infections including syphilis (a sexually transmitted disease), Lyme disease (an infection transmitted through tick bites that may purpose troubles with the coronary heart, joints, and nervous device), or relapsing fever (an contamination transmitted by tick bites that reasons repeated episodes of fever), you can enjoy a response beginning one or hours after receiving your first dose of this medicinal drug and lasting for 12 to 24 hours. Tell your doctor if you enjoy any of the subsequent signs and symptoms: fever, chills, muscle aches, headache, worsening of pores and skin sores, fast heartbeat, fast respiratory, and flushing.

WHEN ANTIBIOTICS ARE NEEDED

Antibiotics can be used to deal with bacterial infections that:

- are unlikely to solve without antibiotics

- may want to take too long to clear without remedy

- convey a hazard of more critical headaches

- may want to infect others

You may nevertheless be infectious after beginning a path of antibiotics. Depending at the infection and the way it's handled, it can take between 48 hours and

14 days to forestall being infectious. Ask a GP or pharmacist for advice. People at an excessive hazard of contamination will also be given antibiotics as a precaution, called antibiotic prophylaxis. Read extra approximately when antibiotics are used and why antibiotics are not automatically used to deal with infections.

HOW TO TAKE ANTIBIOTICS

Take antibiotics as directed on the packet or the affected person statistics leaflet that incorporates the medication, or as told with the aid of your GP or pharmacist. Antibiotics can come as:

- Capsules, drugs or a liquid which you drink – these can be used to treat most forms of mild to moderate infections in the body

- Lotions, lotions, sprays and drops – those are frequently used to deal with pores and skin infections and eye or ear infections

- Injections – these can be given as an injection or thru a drip directly into the blood or muscle, and are used for more severe infections

SIDE CONSEQUENCES OF ANTIBIOTICS

As with any medicinal drug, antibiotics can cause side outcomes. Most antibiotics do no longer cause issues if they may be used properly and critical facet effects are uncommon. The commonplace aspect consequences include:

- being unwell
- feeling ill
- Bloating and indigestion
- diarrhoea

Some human beings might also have an allergy to antibiotics, particularly penicillin and any other form of antibiotic known as cephalosporins. In very rare cases, this can lead to a serious hypersensitive reaction (anaphylaxis) that is a scientific emergency. Call 999 or go to A&E now if:

- You get a skin rash which can include itchy, red, swollen, blistered or peeling skin

- You're wheezing

- You get tightness within the chest or throat

- You have trouble respiratory or speaking

- Your mouth, face, lips, tongue or throat start swelling you may be having a critical allergic reaction and may want on the spot remedy in medical institution. Read greater approximately the facet consequences of antibiotics.

CONSIDERATIONS AND INTERACTIONS

Some antibiotics aren't appropriate for human beings with certain clinical issues, or ladies who're pregnant or breastfeeding. Tell your healthcare expert in case you're pregnant or breastfeeding so one can prescribe the most appropriate antibiotic for you. Only ever take antibiotics prescribed for you – never "borrow" them from a chum or family member. Some antibiotics do not mix nicely with different medicines, such as the contraceptive pill and alcohol.

Read the records leaflet that comes along with your medicine cautiously and talk any issues with your pharmacist or GP. Read greater approximately how antibiotics engage with other drugs.

TYPES OF ANTIBIOTICS

There are hundreds of different types of antibiotics, but most of them can be categorized into 6 companies.

•	Penicillins (along with penicillin, amoxicillin, co-amoxiclav, flucloxacillin and phenoxymethylpenicillin) – broadly used to treat a spread of infections, which includes pores and skin infections, chest infections and urinary tract infections

•	Cephalosporins (along with cefalexin) – used to deal with a huge variety of infections, however a few are also effective for treating extra severe infections, along with sepsis and meningitis

•   Aminoglycosides (together with gentamicin and tobramycin) – generally tend to best be utilized in medical institution to deal with very extreme illnesses which include sepsis, as they could cause critical aspect consequences, consisting of hearing loss and kidney damage; they're typically given by way of injection, however may be given as drops for some ear or eye infections

•   Tetracyclines (which includes tetracycline, doxycycline

and lymecycline) – may be used to deal with a huge variety of infections, however are typically used to deal with zits and a pores and skin condition called rosacea

- Macrolides (which include azithromycin, erythromycin and clarithromycin) – may be especially useful for treating lung and chest infections, or as an opportunity for human beings with a penicillin allergic reaction, or to deal with penicillin-resistant lines of bacteria

- Fluoroquinolones (together with ciprofloxacin and levofloxacin) – are wide-spectrum antibiotics that had been once used to treat a extensive variety of infections, specifically respiration and urinary tract infections; these antibiotics are now not used robotically because of the threat of serious side consequences Other antibiotics consist of chloramphenicol (used for eye and ear infections), fusidic acid (used for pores and skin and eye infections), and nitrofurantoin and trimethoprim (used for urinary tract infections).

WHAT IS PENICILLIN MAXIMUM COMMONLY USED FOR?

Penicillin antibiotic remedy is specifically used to treat bacterial infections caused by both Gram-positive micro organisms, which include:

- Skin infections

- Dental infections

- Ear, nose, and throat infections

- Pneumonia and other respiratory tract infections

- Urinary tract infections

- Sexually transmitted infections
- Brain and spinal cord infections together with meningitis
- Rheumatic fever (a problem of untreated strep throat or scarlet fever) Using antibiotics, which include treatment with penicillins, isn't always effective against the commonplace cold, flu, COVID-19, and other viral infections. Taking antibiotics while they are not wanted (for infections resulting from viruses) can result in the emergence of antibiotic-resistant micro organism.

Antibiotic resistance is a risk to the effectiveness of antibiotics which could store lives. To save you penicillin resistance, you should simplest take penicillin V, penicillin G, and other penicillin antibiotics while they are prescribed by a healthcare expert to treat a confirmed or suspected bacterial contamination.

WHAT ARE THE SIDE EFFECTS OF PENICILLIN?

Common aspect results of penicillin encompass stomach cramps and pain, nausea, vomiting, and diarrhea. When penicillin is run intravenously or intramuscularly, it is able to reason transient aspect effects which include redness, pain, and swelling on the injection website. You ought to inform your doctor if those aspect consequences are severe or do now not leave after a few days. Less usually, taking penicillin can motive greater severe side outcomes, along with both a yeast infection (vaginal itching and/or discharge) and oral thrush (white patches inside the mouth and throat). In uncommon instances, a transient facet effect causing a black furry tongue can manifest. Though this could appearance very worrisome, it normally is going away after completing the antibiotic. Call your medical doctor at once if you increase those unfavourable consequences. Stevens-Johnson syndrome (SJS) is a totally rare but serious side impact of penicillin. Let your doctor realize and are looking for medical care at once in case you word blistering pores and skin rash, fever, and muscle aches. Clostridium difficile (C. Diff) contamination is every other rare facet impact that is very contagious and can purpose severe complications if not treated directly. This isn't always specific for penicillins. All antibiotics can doubtlessly cause C.Diff infections. Inform your doctor if you broaden severe belly cramps, intense diarrhea, and fever.

WHAT IS A PENICILLIN ALLERGY?

Some people are allergic to penicillin and might expand severe symptoms of an allergy, which includes skin rash, hives, itching, trouble respiration, swelling of the lips or throat, or hoarseness. You should are trying to find immediately scientific attention in case you expand signs and signs of an allergic reaction after taking penicillin. It is really worth noting that whilst 10% of the US populace reports having a hypersensitive reaction to penicillin within the past, simplest 1% of humans have a true penicillin hypersensitive reaction. People with a penicillin allergy can be prescribed other non-penicillin huge-spectrum antibiotics to treat bacterial infections

GENERAL PRESCRIBING STANDARDS FOR YOUNGSTERS

Prescribing of any drug in kids requires very cautious consideration of age, weight, and pharmacologic elements of the drug in query. Children are in particular vulnerable to unfavorable drug reactions and dosing errors, and this is compounded by way of the dearth of paediatric labelling info for many not unusual prescribed drugs. The pharmacokinetics and pharmacodynamics of a given drug is often very one of a kind in kids, in comparison to adults, and may vary notably relying on the kid's age and degree of improvement. In preferred, drug dosing in children ought to be weight-based totally (mg/kg), even though note that the recommended weight-based totally dose may vary in step with age (for instance, the 1/2-existence of many tablets is extended in younger toddlers, in comparison to older children, resulting in decrease endorsed doses consistent with weight). Also, maximum dose limits must be taken under consideration and for most pills the adult most dose need to not be surpassed.

Paediatric dosing tables are included in these guidelines to help decide the most suitable dose while antibiotics are required in youngsters. However, these tables do not update medical acumen, and doses have to be adjusted hence if the kid's weight or developmental stage is considered to be outside of the typical values. If in doubt, use the weight-based dosing covered within the tables. Usually, remedy for children beneath 3months (12 weeks) isn't always initiated in the network setting. However, dosing records for this age organization may be covered here for statistics functions handiest and must now not be interpreted as a recommendation to prescribe.

THE END

Made in the USA
Columbia, SC
06 July 2025